Sustainable Routine Stories for a Better Future

Catalogação na Fonte
Elaborado por: Josefina A. S. Guedes
Bibliotecária CRB 9/870

S964s
2023

Sustainable routine stories for a better future / Maria Helena Ferrari [org.] – 1 ed. – Curitiba : Appris, 2023.
155 p. ; 21 cm

Inclui referências.
ISBN: 978-65-250-5326-4

1. Abordagem interdisciplinar do conhecimento. 2. Escrita criativa 3. Sustentabilidade. 4. Língua inglesa. I. Título.

CDD – 370.1

Livro de acordo com a normalização técnica da ABNT

Editora e Livraria Appris Ltda.
Av. Manoel Ribas, 2265 – Mercês
Curitiba/PR – CEP: 80810-002
Tel. (41) 3156 - 4731
www.editoraappris.com.br

Printed in Brazil
Impresso no Brasil

Maria Helena Ferrari
(Organizadora)
Doutoranda no Programa de Pós-Graduação Stricto Sensu
em Educação da Unesp-Marília

Sustainable Routine Stories for a Better Future

Appris editora

Coletânea de textos do Curso Técnico em Informática Integrado ao Ensino Médio do Instituto Federal de Rondônia *Campus* Vilhena 2022

INTRODUÇÃO

Diante da Política de Internacionalização que o Instituto Federal de Rondônia têm adotado para alunos e servidores, em 2021, a Pró-Reitoria de Pesquisa lançou o Edital nº 05/2021 que tratou da Seleção de professores para o Programa de Formação: Educadores do Século 21 em parceria com a faculdade de Ciências Aplicadas de Tampere na Finlândia. O edital assegura a Formação in Locu na faculdade e no retorno, os participantes devem aplicar as experiências com a comunidade em que **estão inseridos**.

Portanto, este trabalho é o resultado de uma das ações do projeto de pesquisa Intitulado Aplicabilidade Transdisciplinar dos Paradigmas dos Objetivos de Desenvolvimento Sustentáveis- Agenda 2030 cujo objetivo é aplicar metodologias ativas baseada em problemas visando promover nos alunos, aprendizagem crítica, tornando-os capazes de transformar a sociedade em busca de qualidade de vida sustentável com foco nos Objetivos de Desenvolvimento Sustentáveis (ODS 2030).

A escolha pela Agenda 2030 deu-se por entender a diferença significativa que o tema pode proporcionar ao estudante para expandir sua compreensão sobre o assunto e colocar em prática o que aprendem em sala de aula.

Neste exemplar, será apresentada a vocês a sequência didática aplicada a turmas dos Cursos Técnicos em Informática, Eletromecânica e Edificações Integrados ao Ensino Médio, que culminou na coletânea dos textos apresentado nesse volume.

SUMÁRIO

SEQUÊNCIA DIDÁTICA

Etapa 01:

Objetivo: Conhecer a ideologia da existência dos objetivos dedesenvolvimento sustentáveis

Material de apoio: computador, internet, material impresso, cartolina e pincéis, conta Google

Duração da etapa: 06 aulas

Estratégiadeaprendizagem: avaliação colaborativa por meio estações de trabalho;

Encontro 01:

Duração: 2 horas-aula. Planejamento da aula:

Nesta atividade, o(a) professor(a) poderá – conforme a realidade da sala de aula – realizar a atividade diagnóstica em material impresso ou cadastrar a atividade diagnóstica no google formulários e gerar endereço eletrônico compartilhável para disponibilizar aos(às) alunos(as).

No primeiro encontro, o(a) professor(a) por meio de diálogo com os(as) alunos(as) apresenta a atividade diagnóstica individual para coletar os conhecimentos prévios dos(as)estudantessobre"Objetivosde desenvolvimentosustentável". https://forms.gle/Y2HfNtnukPx2SLdY9

Nessa atividade, será explicado o objetivo da avaliação, e a sua importância para a seleção de estratégias de ensino mais adequadas ao perfil da turma. Seguem abaixo as questões propostas aos(às) alunos(as):

Apresentação do Tema:

A apresentação do tema foi feita por meio do Vídeo "A maior lição do mundo". Os alunos assistiram ao vídeo por três vezes.

Na primeira vez, foram instruídos para assistirem ao vídeo sem legenda, deveriam prestarem a atenção nos elementos tipográficos e sequência das cenas, na segunda vez: deveriam tomar notas das palavras familiares a eles, sendo cognatas ou outras no caderno; na terceira vez: os alunos assistiram o vídeo com os subtítulos ligados para conferirem as palavras anotadas;

Nesta atividade, o(a) professor(a) poderá – explorar além do conteúdo dos Objetivos Sustáveis, palavras cognatas, prefixos e sufixos, inferências e deduções instigando o aluno a refletir sobre a realidade na comunidade em que vive. O material utilizado foi o vídeodisponível em https://www.youtube.com/watch? v=HW76iO-Q7qVQ sobre os 17 objetivos sustentáveis.

Após assistirem ao vídeo e discutirem sobre os objetivos e vocabulário técnico os alunos foram instruídos a formarem estações de trabalho. Após formarem as estações, cada grupo escolheu um dos 17 objetivos e diante da decisão devem inserir as palavras ou expressões apresentadas no vídeo. Segue o Link de acesso a atividade . https://jamboard.google.com/d/1_njN1ErmUCn0jpqU_w- QYShg_5MFlT-mq8QN4rubxE7w/edit?usp=sharing e a ser feito pelo smartphone.

Encontro 03:

Duração: 2 horas-aula.

Planejamento da aula:

Nesta aula faremos a avaliação colaborativa por meio das estações de trabalho por diferentes competências. O professor cria as estações de trabalho de acordo o número de alunos e espaço para realização da atividade e modo que todos os alunos passem por todas as estações.

Estação 01: JIGSAW

Nível fácil

Objetivo: reconhecer os objetivos de desenvolvimento sustentável pelas imagens e colocá-los na ordem correta

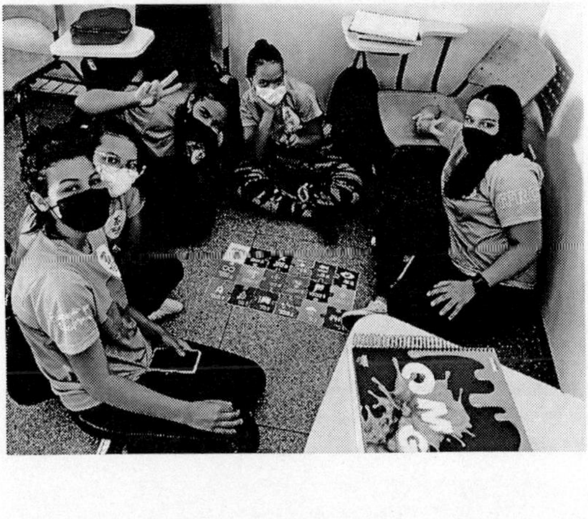

Etapa 02: Game MATCH

Objetivo: Produzir textos narrativo-descritivos que abordem problemáticas apresentadas na Agenda 2030 no dia-a-dia dos cidadãos.

Materiais de apoio: computador, internet.

Duração da etapa: 06 aulas

Estratégia de aprendizagem:

Avaliação Individual por meio de feedbacks.

Estação 04: Desenho

Nível Difícil

Objetivo: Representar os 17 Objetivos de Desenvolvimento Sustentáveis por meio de habilidades artísticas.

Etapa 02:

Objetivo: Produzir textos narrativo-descritivos que abordem problemáticas apresentadas na Agenda 2030 no dia-a-dia dos cidadãos.

Materiais de apoio: computador, internet.

Duração da etapa: 06 aulas

Estratégia de aprendizagem:

Avaliação Individual por meio de feedbacks.

PLANTS IN KATE´S LIFE

Aline Noemerg Grey

On a morning walk the young woman, called Kate stops in front of a large market to have her breakfast.

While passing between the stands she sees a small cactus and becomes delighted with that small fragile plant with its thorns defending its small buds. For some reason, this encouraged her to buy it. So the young woman bought her first small plant and was willing to take care of it.

Arriving home, the young woman goes straight to the window of her house, where she leaves her small plant there to catch the sun, so that her small cactus can absorb energy by the sun´s lights.

When night comes, she waters the cactus, because at night it takes time to absorb the water and this makes the plant stay hydrated longer.

The young lady started to do this every night, but she started to feel that her plant was getting lonely, so on weekend she decided to go back to the fair. She decided to go back to the market and buy new plants, this reminded her of her mother and brought to her heart happiness.

In a few months her house was full of plants, which brought a better quality of life, mentally, and her health. She started to eat more vegetables and fruits, and in the future, she plans to move to a bigger place, so that she can have more space for her cultivation, and in the future live her old age beside her plants.

BONNEY THE FISHER

Asafe Henrique Custodio Barbosa

Bonney is at the ranch. She enjoys the laidback life of the farm. One of her best hobbies at the farm is fishing.

Some days ago, she went fishing and found trash in the water. She noticed that when the trash was crossing the river there were several trucks unloading the trash. She became very angry. She stopped and tried to help nature, but they didn't listen to her.

Her dream is to see the river clean again, but she doesn't know how to do it. She thinks she is doing her part to make people and friends aware of not throwing trash in the rivers.

Over time, she cleans up her river and makes a recycling system in her river so that all the trash can be reused, and she has started making tools to grow crops on her ranch from the recycled trash.

She saw that she could help people in need with recycling. People in town thought her initiative was very good, so they helped her.

All the neighbors on her ranch started doing the same and realized that recycling really worked. At the end of the month, they had a display of artwork made from plastic bottles.

HOPE ROUTINE IN ITALY

Élen Nayara de Souza Albuquerque

Hope had just entered the university of her dreams, in the city of her dreams. She couldn't believe that she had gotten a scholarship to study photography in Florence. Since she was 16, her dream had been to visit Italy and go to art school. But now, she was going to live there for about six months. When she got the news she left behind her nutrition college, which her parents had convinced her to study. At that moment, nothing was going to change her mind. Hope had the conviction that this was what she wanted.

The great day arrived. Watching the clouds through her airplane window, she wondered what her life would be like in a country far away from everything she would ever know. About how disappointed her father would be with her, or how terribly angry her mother would be at her. Even more, when she just picked up her things while they were at work and said goodbye through a single voice message she had sent to her father's cell phone

She stopped and thought: Now, Hope. Let's not think about bad things. You're about to realize your greatest dream. She said with the intention of convincing herself. After arriving in Florence Hope is amazed at the beauty of the city, the only thing that bothered her was the bad smell of garbage. The smell that she would experience day after day since she was about to start her life in that city.

Some months passed and Hope had gotten used to the place, she lived in a small apartment, all in a woody tone. In the kitchen window there was a small mother-in-law's tongue, a plant that helped to improve the air quality. So many days had passed that she didn't even realize she had already begun to use the Italian's habits. For example in the way of dressing, and cooking. She was also speaking Italian easily.

Hope spent from Monday through Friday practicing her Italian. It became routine for her to take at least one lesson every night on one of her apps. It was always the same, she would wake up around six in the morning, put on her sports clothes, and go for a run. When she returned home she would have her breakfast and get ready to go to her college. It was a rather dull routine, but she loved every second of it.

On Saturdays, she would hang out with some of her friends who were also exchange students, Carolina a North American who was doing fashion design and Howard a Norwegian who had gone to study art history. Are you ready, Hope, to go home? Howard asked. Not really, I'm going to miss this place and its magic. The way people care for the environment, the way teachers teach in the schools. I will really miss this place, the delicious food, and especially the gelato. I can't believe it's been six months already.

AMANDA'S SUSTAINABLE ROUTINE

Gabrielly Serafim de Oliveira

Amanda is 14 years old. She has had a sustainable routine since she was very young. Every morning, she wakes up and eats a healthy snack. Before going to school, she organizes all the garbage. Each one in its proper place.

She goes to school by bike and takes her own snack that she prepares herself. She also takes her own water bottle. So, she doesn't use disposable glasses.

One interesting thing is that her school supplies are made of recycled paper. She takes good care of her backpack so that she doesn't have to buy one every year.

When Amanda gets home, after school, she waters her plants and cleans the house with soap and vinegar. After cleaning everything, she once a month, she checks if there's anything she doesn't use anymore to donate.

Then, in regular days she takes a short shower and brushes her teeth with her homemade toothpaste and wooden brush and then turns off all the lights and sleep.

GO GREEN

Jennifer Rebeca Mateus Nascimento

Jennifer was born in Rondônia. When she was 16, she moved to a French countryside, to Annecy, a very beautiful and lively city. Annecy landscapes are full of rivers, and it has many flowers in spring. She loves to walk around town with her dog, keny, in the public squares, enjoying the view and thinking about how she can introduce her little sustainability project ideas that she has from her college into that small town. Lately, she has been very busy with the preparations for her trip to the country of her favorite book "The light through the window".

Now, she can make this dream come true through her 8-day trip to Oxonia, in England. She will be there for the immerse essay competition.

As usual, the day starts very busy. Jennifer wakes up at 7. She takes a quick shower to avoid wasting too much water and starts the day well because she lives alone. She, routinely talks to the plants that she has in many parts of her house, mainly at the windows to improve air quality, and especially the English ivy because they look beautiful fallen on the windows and wall.

Jennifer does not like to print papers to find a waste of paper because it only pollutes the environment if disposed incorrectly. Unfortunately, it is what many people do. So instead of printing paper unconsciously, she always uses a notebook even to write down what she would need to print. After that, she no longer uses the notebook. She discards it in a proper place, but now she needs to print the poster of his sustainability project that she will present at a lecture. She received the opportunity to give at Oxford College a presentation about the importance

As usual, the day starts very busy. Jennifer wakes up at 7. She takes a quick shower to avoid wasting too much water and starts the day well because she lives alone. She routinely talks to the plants that she has in many parts of her house mainly at the windows to improve air quality, and especially the English ivy because they look beautiful fallen on the windows. Jennifer does not like to print papers to find a waste of paper because it only pollutes the environment if disposed incorrectly.

Unfortunately, it is what many people do. So instead of printing paper unconsciously, she always uses a notebook even to write down what she would need to print. After that, she no longer uses the notebook. She discards it in a proper place, but now she needs to print the poster of his sustainability project that she will present at a lecture. She received the opportunity to give at Oxford College a presentation about the importance

Months later, on the day of the presentation, she won the UN environmental award, for several of her ideas that helped the world in relation to environmental pollution and the preservation of the environment.

HEALTHY ROUTINE

Nicoly Eduarda Pacini dos Santos

Olivia is a 17-year-old girl. She decided she would eat less meat. Olivia has always been concerned about her health and when she discovered the harm of red meat she decided to become a vegetarian.

Olivia became increasingly interested in vegetarianism and with that, she began to study the harm of meat in our environment. So, she starts a project on social media to try to make people aware of the excessive consumption of meat and encouraged people to study this subject.

On social media, she also posts about her healthy routine once a week and participates in several recycling projects throughout the year. One of them is to collect garbage from the rivers and turn them into toys to donate.

Another action is that she works at an NGO that rescues animals. She takes in stray animals, and feeds, and cares for them until they find a loving owner.

Olivia dreams of a world where everyone starts helping the environment, and she tries to convince as many people as she can.

THE SUSTAINABLE MAN

Pedro Henrique Baltazar da Silva

Once upon a time there was a boy named Vitor who in his childhood didn't know what agenda 2030 was. When he learned about it in his school, he started to plan his routine according to the goals of agenda 2030 to become a reality in his town.

Now he is 25 years old man. He works as a Civil Police Officer. His friends and he save R$10,000 reais. With this money he starts to invest in the first objective "ZERO HUNGER".

Vitor buys several basic food baskets. He goes to the poorest neighborhoods and distributes the food once a month.

He sees the happiness of those people and realizes that it was good for him too. So he invests in the objective of "QUALITY EDUCATION".

To fulfill this task, he goes to schools to give lectures talking about how education can lead you to have a good job and a good preparation for life.

Vitor barely finishes the lectures and has already started planning the goal "LIVE ON LAND" where he surrounds his city in search of garbage, and waste pollutants that cause harm to nature.

Finally, he finishes with the goals his boss promotes him to sergeant and the mayor gives him the key to the city along with the construction of his own statue and thus he became known as the Sustainable Man.

LEARNING SUSTAINABILITY

Sidnei Salustiano Paulino Júnior

One day, a girl named Laura was walking on the sidewalk until she saw a billboard about the 2030 agenda.

Laura thought it was stupid because her city was well structured and she hardly saw any problems in it. She didn't understand the reason for that billboard.

She continued walking and soon after she saw several protests about the pollution. She passed straight through, but this time that was in her mind, is the world really in danger?

Two weeks later, Laura kept walking, this time she was getting far from her city when she realized that she was in an industrial area. She started to feel a strange smell. It was the pollution from the industries.

Laura, slowly, started to remember the protests, the agenda 2030.

On the way back home, she wondered how that pollution could harm the population, and she slowly became worried. She decided to develop a project, against pollution, with the objective of starting to use clean energy.

Gradually her project was gaining members and having influence in her community.

On a Friday afternoon, Laura was protesting in front of the government headquarters. In the event she is expected to give a speech:

"-I have a dream, that one day, we will use only clean energy. Thus decreasing pollution, increasing life expectancy of our planet, making a good place for land and sea living beings. We cannot stand by and watch the world destroy itself, we must act."

That afternoon many people listened to Laura's speech, supporting her project.

That day was historic, Laura got an agreement with the government to reduce the pollution from the factories. There were still many things missing to make the planet a better place. So, Laura, started working on all the 2030 goals, striving to contribute to the environment.

RUSSO'S IDEOLOGY FOR A SUSTAINABLE WORLD

Thauan Wictor Muniz Alves

On a fine day, around 5:00 in the morning a smart boy gets up for his morning cardio, which he does at least twice a week. During his cardio, while listening to his podcast, about sustainable living and how to put it into practice. He realizes how different our world is from what it was meant to be, for example, while the speaker approached the goal "zero hunger".

On his walk, he realizes how many people live on the street, and don't know if they will have their daily food and bath.

The guy became intrigued by what he had heard in the podcast. He tries to know how to have a better, more sustainable world. So that, humanity will not perish. He begins to realize how difficult this job is and finally gives up because it is a very difficult task to do. There is no way to do it alone and no friend would help him.

Instead of wanting to help. No!

They made him give up faster by saying:

Dude, give up, there are millions of people going hungry. do you think you can eradicate their hunger?

You never will!

Finally, he reached one of his conclusions. The problem is not hunger, but the men themselves (humanity), with this bitter heart, where we cannot blossom a flower of charity and love for others. It is difficult to have a sustainable planet! We must first put in our mind "if we continue like this, there will be no tomorrow! We will die before we see such a "better world".

We have to start building a better world, by having fair public policies to improve our education, which today is precarious. Students with incredible minds, study in schools with the roof falling on their heads, going hungry to school, and being exposed to a hot stuffy room where we have more than 40 students, with incredible ideas with incredible proposals.

Because of the lack of a chance at school, students don't get what they want, and unfortunately. We lose incredible minds to the world of crime or even wasted where they should be using their knowledge somewhere else doing what's not going to make a difference to the world.

So, in conclusion, we need to make human beings aware of the need to be more compassionate with their neighbor. Their neighbor may even be their son, their grandson, their great-grandson, and their close relatives.

Recycling, and knowing how to use drinking water, are also very important because there are

places that suffer from garbage and the waste of fresh drinking water.

Because of the waste of water tthat kills marine life, the waste of human waste such as bags, contaminated sewage water, and oil that could not be collected and is wasted in the sea.

There are studies that in a few years there will be more garbage in the sea than marine life but we know that marine life is also very important for our ecosystem.

CARLIN AND EQUALITY

Yan Fernando Lobato

Carlin is a kind boy, he always helps those in need, donating to institutions, homeless people, people who need money, etc... He loves to make people happy. He is always there for all people, but doesn't care much for him, despite his poor health, many people had prejudices against him. One day Carlin, as usual, went to help a needy person on the street, but when he went to deliver a hot lunch to the beggar. Carlin realized that the beggar was ignoring him because of his appearance and color. After that, he left his lunch on the floor and left. After two weeks being sad in his bed, he decided to change his appearance, not to please others but to please himself. Three months after this event he was at the peak of his health and appearance. As usual, Carlin went to help people in need, but walking across the street he saw two men who were fighting. He tried to stop the fight, but the realized that one of them was a racist. Carlin tried to call the authorities. At first, the racist was trying to resist. But after, he realized that Carlin was big and strong, he became afraid and left him. The person that the racist was fighting was a needy person. Carlin helped him, but something different happened, the needy promised Carlin to be like him and help as many people as he could. An attitude can change people.

LUCY' SUSTAINABLE ROUTINE

Dalila Elena Pelegrinello

Lucy is 20 years old and studies marine biology with the intention of studying and helping marine life become part of her routine.

Her routine starts at 6:00 a.m. when she wakes up to get ready to go college. At 6:30 a.m. Lucy has breakfast and then leaves so she won't be late.

Her class starts at 7:20 so Lucy spends her day studying, 13:30 is time for lunch, Lucy prefers to stick to healthier foods like fruits and vegetables that she brings from home. After studying Lucy takes care of her plants dayle and walks a little before going to sleep at 20:00 PM.

APPLE TREE

Júlia Eduarda M. Oliveira

Once upon a time there was a child who was not much different from the others, she used to play, eat hamburgers and sleep long hours, just like it is normal nowadays. She didn't see any problems living like that, until her friend Robin visited her, bringing with her a small plant in her arms that was a little hard to identify. Perplexed by the scene, she wondered:

- Why do you bring a plant to my house on foot, when I could come by car without it?

- Because I don't mind walking so much, so don't pollute the air, and the little plant is just to help you.

- How?

- Plant it in your yard and it will do its part in purifying your air.

After this dialogue the child started to test her friend's way of life and after months she was already not letting go of her new routine that before she thought she would never want.

Today she left her house early walking to the institute where she studies, smiling with her eyes brighting, with a pet bottle in her hands that she just recycled!

She came home for lunch and picked some vegetables from her own garden to eat and watered her beloved friend's plant, an apple tree seedling, which she can now identify.

Until recently she thought that living sustainably was silly, but she has learned that it does take work. There is no way. You have to give up some things, change your routine a little, and give up your laziness.

But the good thing is that you can make it all fun. That feeling of change is sometimes good to give you an extra boost, and you will sleep peacefully knowing that you are helping to build a better world. Robin taught this to his friend and now he can't live without this learning.

MILENA AND HER WEEK'S ROUTINE

Ana Carolina Leopoldo Dos Santos

Milena is 16 years old and loves vegetarian food. Her favorite food is eggplant strogonoff. She also loves guacamole an appetizer made with avocado, tomato, and some spices. She eats guacamole every day in the morning or afternoon when she has avocado in her house.

Milena is different from other teenagers of her age. She loves to go to school by bike while the other teenagers love to go by car or by motorcycle. She tries to talk them to go once a day by bike to help the environment, but they don't even care about what she says.

Milena loves nature, so when she is not studying she does gardening with her mother. She also loves to go for walks in the countryside together with her family or going to the farm they have. There she feels it is her place with fauna and flora with waterfalls and rivers, a calm and peaceful place.

She loves adventure, she also loves animals and loves sports, especially volleyball. She loves to smell the scents of the flowers in the garden at her school. Her favorite place at school besides the garden is the sports court and the library, where there are many interesting books for her to read about the environment. In the very near future she wants to write a book about environmental sustainability.

MIGUEL'S SUSTAINABLE LIFE

Pedro Noemerg de Andrade

Hello, I am here to talk about Miguel's sustainable life. Miguel is 28 years old, lives in a small town, and lives alone in his house. Every morning he gets up and rides his bicycle to work, and for lunch, he eats vegetables and fruit he grows at home and brings in his lunchbox. Every afternoon he tends his vegetable garden, and every Sunday he goes to an open market to sell some vegetables and fruits that he has produced and left over from his garden.

WHAT IF IT WERE DIFFERENT

Wanda Pereira Da Costa

The year is 2015, when the UN created a plan to transform the world by developing a more sustainable world. The idea initially seemed unattainable, targeting the year 2030 totally different from what was predicted.

Everyone saw 2030 as a year when global pollution and other environmental disasters would be totally aggravated to such an extreme level that it would be almost impossible to continue living on planet Earth. But the United Nations created an agenda with sustainability acts being 17.

Taking care of the planet and its inhabitants was important at that time, and everyone collaborated by having healthy habits and helping to combat any means of pollution to the world. They preserved water, no longer wasted food, the world was egalitarian with everyone, no one having more rights than the other, they used less means of transportation that release polluting gases, and reforested more, disposed of solid or organic waste properly, recycled, and each time there was a different way to reuse.

Education changed completely, now having revolutionary children who never stopped thinking about the future and different ways to preserve life around them. People helping each other, and because they now had healthy habits, they have a longer life rate. Life was completely different and the air was cleaner, the Earth had never been so wonderful as it was in those days. After 15 years of the inauguration of the 2030 agenda, the planet was the way they had envisioned it back in the past with a completely better life and was getting better and better, developing better and better habits.

Life was completely different and the air was cleaner, the Earth had never been so wonderful as it was in those days. After 15 years of the inauguration of the 2030 agenda, the planet was the way they had envisioned it back in the past with a completely better life and was getting better and better, developing better and better habits.

Unfortunately the story is fictional, possibly someone had this dream that the world could go forward full steam ahead, with everyone in the fight to save the planet, but it is not quite like that, some nations can follow to the letter each sustainable goal, but the rest of the world would not go so fast in this walk and who knows until everyone is practicing to the letter the agenda the world has not been extinguished.

But don't stop trying yet to change the course of things, who knows, maybe the dream will come true.

EFFORTS FOR A BETTER FUTURE

Lemuel Lemes

Cleiton boy likes all people, always dreaming and having hard days but always dreaming of finding an opportunity to show his abilities. He is always smiling, making people happy with his happy way of being, always looking for something new to know. He has a kind of complicated life because he lives alone. He doesn't have many friends because he is kind of lonely but he is very skilled in mechanics. His biggest dream is to become a great professional mechanic. The first step to be taken is that he needs to show his abilities to companies that are very famous and extremely sought after by the public.

So, he started searching for some opportunity. On an ordinary day, he was passing in front of one of the most famous workshops in the world when he saw a poster that said "we are opening a selection to choose only one mechanic to travel out of the country to attends a professional course in mechanical engineering with the value of 250 $" Cleitinho, then, saw this and became very happy because it was a golden opportunity. Immediately, he went to make his registration. So, after his registration was done, he had a week to prepare his interview. He studied non-stop, so he can gain more knowledge and be prepared for this selection.

After a week he continued focused to get there. There were many competitors. So the selection started. There would have two phases. The first phase was a test to check his knowledge about mechanical engineering, thus ending the 1st phase, only he and one more person and so, they left for the second phase which was exactly a mechanical operation to be done in about 5 hours. After it was over, the result came out after a lot of effort, he got the place for the course and so he took his things getting everything right and so he went with this chance he will be able to have a better life and achieve a sustainable routine.

MURILO'S SUSTAINABLE LIFE

Eduardo Fernandes

Hello, today I came to tell you a story about sustainability. It all starts at Murilo's school, where the teacher explains what the goals of sustainable development are and explains its importance.

And when Murilo gets home he realizes that he needs to make some sustainable changes. So, the next day he wakes up and instead of driving to school he decides to ride his bike to get some exercise and not pollute the air. Then, he comes home for lunch, and when he arrives instead of eating bad food he eats greens and vegetables. He also decides instead of throwing his garbage in any garbage can, he decides to throw it in recyclable garbage cans. And he also started to plant some fruits and vegetables to have a more sustainable and healthy life.

He also starts recycling his electronic waste, and at the end of the day he instead of sleeping late, he sleeps earlier to get a healthier night's sleep. And the next day he wakes up, eats breakfast, and rides his bike to school again. When he gets to school he tells his classmates and his teacher how his life has improved with the sustainable changes he has made to his routine. Then, his classmates decide to make the same sustainable changes as Murilo, because they notice that Murilo is happier and healthier. In this way, Murilo and his friends began to have a sustainable routine.

SUSTAINABLE ACTION FOR A BETTER FUTURE

Erick Brayan Boscardim Barbosa

Pedro is a twenty-one-year-old young man who lives alone. He is quiet and likes to be in contact with nature. When he wakes up in the morning, he has his breakfast and walks with his dogs in the park. He has many friends, but he likes to spend most of his time alone in contact with nature.

Once a month, he goes to the supermarket named Good Success of his neighborhood, to buy and exchange "left over", as he has a good amount of money "left over" with that money makes a donation to charity every month.

In the afternoon, he goes to work and doesn't come back until after 5 pm. He works three days a week. On the other days he participates in environmental projects at his college, such as recycling projects or even planting tree seedlings or taking garbage out of the rivers.

When he gets home, he usually does his college activities, so he doesn't let it accumulate.

At night, he takes some time for himself, for being very hygienic, like nights he cleans his house and separates his own trash. After he finishes his chores, he goes to the gym, he usually goes every day of the week, comes back twenty-two hours and goes to bed to rest from his long productive day.

SARAH AND HER EVOLUTION

Patricia de Oliveira Silva

Sarah maintains a healthy and sustainable life, she is Always looking to do things in a environmentally friendly way. Every 3 months she donates for charity institutions something she no longer needs. She always tries to turn off the lights when there is no one around and tries not to waste too much water. But she was like that, Sarah was already a very busy person and she didn't care about these things, but she realized that this could change the current situation, and if everyone thought like that, it would be very good.

From Wednesday to Friday Sarah rides her bike to work. Sarah takes advantage of it and improves her health, avoids consuming food that is harmful to health and avoids using straws in restaurants and plastics. In her house she has a separate bag for things that are recyclable.

One day I was in restaurants and saw that a person took 6 straws, went to the square and threw the can and the straw on the floor, she certainly didn't like that attitude and went to get satisfaction, asking why that person had done that. So, beside him there was a trash can off his own. The person left without argument, embarrassed and never did that again. Today he made a partnership with Sarah to one of the biggest sustainability companies, NATURA.

PAUL'S LEARNING

Uédina Clarice B. Da Silva Barros

Hello everybody!

Today I will tell you the story of Paulo. Paulo was a boy who didn't care about the environment until one day, he won a scholarship to Switzerland. The long-awaited day has arrived! Paulo packed his bags for traveling. After a long and anxious trip, Paulo arrives in Bern the capital of Switzerland. He wakes up at 05:00 in the morning and has his breakfast with lots of fruits and carbs. Then, as usual, he takes a walk.

During his walk through the streets a girl calls him:

-Excuse me, young man, do you know the 2030 agenda?

Paulo says: -Not !! Politely.

The girl responds : -That must be why you throw waste on the street.

Said the furious girl and gives him back the packet of candy that Paulo had thrown on road.

Minutes later, the girl turns around and continues walking.

Paulo calls her and asks:

-What is the 2030 agenda? and what is it for?

The girl answered:

- Agenda 20 30 is a global action that was taken in 2015, involving 193 countries including Switzerland and there are 17 goals and its 17 goals are geared towards tackling the world's biggest challenges such as poverty eradication and sustainable development, among others.

Paulo is amazed at how such a small young woman cares more about the environment than he does.

When Paulo arrives at his university, he notices that his classmates are writing an essay on sustainable development. Paulo became very happy because he will write in his text everything he learned this morning.

JOÃO'S STORY

Alisson Dias

There was a boy named João, who loved to fish in a river near his house.

One day he went fishing and realized that something was wrong, the fish in that river had disappeared and he wondered why.

So, he noticed that in the banks of that river there was a lot of plastic and electronic waste. João decided to call his friends, and together to make a demonstration about the situation of that river and guide the disposal of this garbage.

João said:

– Please stop discarding garbage in that river, because we need it for our life.

João urged the residents to change their attitude, as we would have many problems in the future.

And with that the residents changed their attitude and started to recycle this garbage so that it could not be discarded in the river.

In a short time the river was recovering well and the fish reappeared and João returned to fish with his friends.

STUDENT IN A SUSTAINABLE DAY

Natam Menegheti Farias

A certain municipal school student decides to fight to have a better school, his school does not have a good structure nor a good education.

It ends up that this student talks to the school principal to find a speaker to talk a little about agenda 2030.

The principal thinks it is a good idea to give this lecture.

Then, after a few months they find a speaker, this speaker talks about agenda 2030 and the plans for a better world.

After these lectures some students understood that you can not judge a person by gender and the school began to do some activities to improve education, began to give lectures, and study about a building healthier and more sustainable planet.

SUSTAINABILITYY

Sebastião Neto

Imagine a Hollywood movie, about the apocalypse, the end of the world.

Well, you propably saw poverty, misery, wars, or aliens (Hollywood just like aliens and explosions).

This looks impossible right?

But no.

Our planet is going to the end, day by day, and, if we do nothing, eventually the problem won't have a solution.

Today, the United Nations created a plan, the agenda 2030.

What's agenda 2030?

Basically, is a list of goals we need complete before the year 2030.

If you want know what is the goals, one by one, you can search about in internet. Beeing short, the nations, and the population of course, will try reduce the poverty, pollution, corruption and inequality.

PAULO'S SUSTAINABLE ROUTINE

Keila Rodrigues Silva

Paulo is 16 years old, lives in Portugal with his parents, they have a sustainable routine, and are trying to show the rest of the family how important this is and it gets to be fun.

They spend more time together taking care of the garden, they go walking to places and the conversation flows better. Paulo at school tells his classmates how important it is to have a sustainable routine, and explains the importance of this to the world, and a few months after insisting on this subject he has as a result 10 of his classmates decided to try a sustainable routine.

MICKEY AND THE WAY HE HELPS HIS CITY

Janine Rodrigues Schulz

One day Mickey Mouse and his friend Goofy realized that because there was a lot of deforestation near their house the climate and air was so dry, so they got a great idea.

-Mickey said: Goofy, what do you think about planting fruit trees for not only improve the air and climate but also to obtain fruit ?

-Goofy: Mickey hehe, I love your idea, let's call more friends to make this project even bigger. And they went to call some friends, who will ask what the purpose of this is.

- Mickey and Goofy: To help the sustainability of our city.

-His friends: what is sustainability?

-Mickey: What do you mean you don't know? -Goofy: Sustainability involves understanding the impacts of our products and waste, as well as the dependence we have on resources, which is crucial if we are to continue to survive.

His friends understood and were happy to help, they all planted at least 10 trees and from then on they all started to inpire people about sustainability to their community.

RS-2030

Fernanda Gonçalves da Costa

A long time ago there was a planet called RS-57. It was inhabited by little creatures, fairies. These were the little creatures that took good care of their planet.

A little further away, there was another planet called Earth. It was inhabited by human beings who didn't care about the welfare of their planet and only cared about machines and technology.

Humans began to die because of their lack of care for their the planet. Little by little the human race ceased to exist. 35 years later there was no human or animal life on earth. Just one species has still been alive, the cockroaches species.

The fairies from planet RS-571, while sailing through space saw that deplorable planet. And landing there they said :

Enour: This planet is so destroyed it makes me want to cry;

Elin: Yes, and thinking about it now, this planet is huge, there is so much water, and so much green, and those different buildings are exotic;

Enour - What is it? - He says bending down and picking something up - it's disgusting but cute at the same time;

Elin- It's so cute what's the name of this thing?

And further ahead they saw a glass of repellent thrown, and in it was a picture with some insects and among them the one that was in Enour's hand, and it also was written the name of that cute and strange creature.

Elin - It seems that its name is Cockroach. What an ugly name to give to such a cute little animal. I will call it Edoop;

Enour- Cute name, but now we have to look for our brothers, this planet is perfect for us to live here. Thinking about the space wars, our planet is in serious danger;

Elin- We will get them and with your help, we will take care of this planet the way it deserves;

Enour and Elin went back to their planet. They found their 17000 brothers and sisters. They loved the planet and together started to organize things.

Elin- Blue Fairies with their water-maneuvering powers took care of the sea, you will ensure CLEAN WATER AND SANITATION and LIFE BELOW WATER. Besides that, you will ensure a perfect place for the TEDX9 (species of marine life on planet RS-571);

Fairies- We promise;

Enour- Green Fairies with their magical and delicate hands to tinker with flowers, trees, and earth, you will ensure the care and LIFE ON LAND, and ensure a home for every species of PICB7 (name of the species of life on earth of planet RS-571);

Fairies- We promise;

Elin- Yellow Fairies with their powers of light, you will guarantee AFFORDABLE AND CLEAN ENERGY, to ensure light and warmth in the cold winter days for all the little fairies;

Fairies- We promise;

Enour- White Fairies with youbsensible minds, you will take care of PEACE, JUSTICE AND STRONG INSTITUTIONS also take care of DECENT WORK AND ECONOMY

Fairies- We promise;

Elin- Red Fairies and their healing powers and education will take care of GOOD HEALTH AND WELL-BEING and QUALITY EDUCATION;

Fairies- We promise;

Enour- And we will take care of RESPONSIBLE CONSUPTION AND PRODUCTION and GENDER EQUALITY. Fairies we will start our jobs. We will rebuild this world and we will make it a better place.

The fairies begin their work. Days and nights non-stop, cleaning the lakes, rivers and seas. Also working better on the infrastructure, removing garbage, giving a good education with great teachers who were very dedicated to that, with more accessibility for the cities providing drinking water, and accessible renewable energy to all citizens who lived there.

The world is not the same as before. Now it is RS-2030 the planet cleaner and with peace established by the people, everyone lived in harmony.

EDUARDO E ISABELLA'S ROUTINE

Eduardo Souza Chaves
Isabella Caroline Souza Chaves

My name is Eduardo Souza Chaves, I'm 17 years old and I study at the Federal Institute IFRO Campus Vilhena. I go to the gym every day. I exercise and I like to play volleyball. I also like to go out to observe nature. I always play a lot of volleyball with my friends and then we walk along the AABB club.

I use to go for a walk with my family, but sometimes I don't walk because I don't have time. I really enjoy playing games, but I always avoid leaving my cell phone charging to save energy. I help my father take care of the house's finances and I always avoid leaving the lights on for no reason. I like to go out with my friends and family, I also really enjoy camping, and playing games, I always walk to the gym and I go camping with my church people. I have a healthy diet, I eat fruits, I eat well, and I drink a lot of water. Now, my sister is going to present herself and talk about her sustainable actions.

Hi, everyone, I'm Isabella Caroline Souza Chaves I am 17 years old and I study at the Federal Institute IFRO Campus Vilhena, I go to the gym, I exercise and I like to play volleyball. I also like to go for a walk with my dog and my mother, I used to go for a walk every day. But today I don't have much time to go out for a walk. I always help my mother to take care of the plants, my mother loves orchids and here at home there are a lot of them. Every Sunday we take care of the plants here at home. I like to go out with my friends and family, I also really like camping, but there are times when we can't camp in the middle of nature because of deforestation and fires. This weekend I'm going camping with the teenagers from my church. In my free time, I usually eat fruit and drink water, since I was little, I also read a lot. As you can see, our parents, teachers, friends and family are always inserting good actions and things for a better future. What are you doing for you and for your community?

KYRIE'S ROUTINE

João Luiz Rossin

It is a Monday morning, Kyrie wakes up and goes to his usual restaurant for having a cup of coffee. He takes his drink and continues his walking.

Halfway from there, he rents a bicycle instead of using a combustion vehicle. Kyrie arrives at his office, where he works as a programmer, and after a while, he takes a break for a snack, he eats an apple with juice.

The day is ending, and Kyrie's work shift comes to an end. He gets on his bicycle and after a good ride back home, he takes a nice shower, in which he tries to save as much water as possible. After Kyrie getting out of the shower, he usually waters his fruit plants in the garden. He collects some vegetables, which he always uses for his dinner.

Kyrie makes his meal, vegetables, fruits, salads, and he doens't eat much red meat. Finally, he sits on his couch and watches the news on TV. After a break, he brushes his teeth (again saving water) and goes to bed, and this is Kyrie's routine.

MY ROUTINE

Marcus Vinicius

I'm Marcus Vinicius. I go every morning to the Federal Institute of Rondônia. I am a student there. I take a technical course in Computing.

I have a big family. So I have a busy daily routine. I wake up every day at 5:30 am, feed the dog, brush my teeth and pack my school supplies because we are a big family, we have to save costs like not leaving the lights on unnecessarily, checking the water taps, among other things.

Dona Alaide, in this case my mother, taught me that we must maintain good personal and environmental hygiene, so I do a general cleaning in my room every Friday afternoon. Because I'm in high school, I need to dedicate myself a lot to the subjects, so I keep a weekly study routine, every afternoon I reserve a few hours of my time to study, I didn't have this custom in the past.

In addition to studying, I like to practice daily exercises and I mainly recommend this to my friends. It is even necessary to maintain a good diet to have a good health performance, I go to the gym from Monday to Friday, I practice daily runs, because I am an athlete and I need to train for competitions while maintaining good cardiovascular health.

To have a good athletic performance, I have to maintain a good diet with restrictions, so I avoid consuming foods that can harm my health, replacing it with protein foods.

In my academic life, I participate in projects, one of them is Biology, which I learned a lot about anatomy, food. After all, biology is the study of life.

ANA JULIA'S ROUTINE

Kauan G.Ribeiro

Ana Julia always wakes up at 7:00, has breakfast, and gets ready to go to school. Then, she takes her backpack says goodbye to everyone at home, and goes to school. Julia always goes to school by bike because she hates depending on others in addition to being healthy and it is great for the environment. Her class starts at 7:10 and goes until 11:35.

In the afternoon Julia is exhausted, so she has lunch with the family and rests a little. Today she has homework to do, she doesn't like to leave it to the last minute, so she quickly starts doing the exercises. After that Julia is free so she has a moment of relaxation where she chats with friends, practices some sport or interacts with family members. At night she takes a shower and goes to the dinner table. After dinner, she helps her mother to wash the dishes, Julia always watches tv or plays some game on her computer. Right after, she goes to the bathroom to brush her teeth and then goes to the bed to rest because the next day another routine begins.

SUSTAINABLE ROUTINE

Mel Misgrey Prestes

You don't have to make big sacrifices to make a difference. If everyone incorporates simple attitudes in everyday life, it is possible to positively impact the environment and society.

I woke up in the morning, turned off the fan, so as not to consume energy, opened the window to let in the beautiful sunlight, and refresh the environment.

I go to the bathroom to brush my teeth and use the necessary amount of toothpaste (no exaggeration), avoiding leaving the tap on, only when using the water, when taking a shower, using cold water during the summer, so as not to waste water unnecessarily.

When I have my breakfast, I always separate the fruit peels for organic waste; stuck always sorting recyclables....washing the coffee dishes, always reusing the water (if possible), and not forgetting to check all the faucets to make sure they are all properly turned off!

When going to school, I always pay attention to the trash cans and their colors, so that all discards are properly disposed of. I always advise people to participate in environmental awareness projects! so that they can get an idea and share the message so that people can also collaborate with the environment.

Coming home from school, I always make the most of the sunlight, leaving clothes and shoes to get some fresh air.

When preparing lunch, I properly dispose of the garbage (reuse of packaging, separation for recycling, donation of items to social projects or engaged cooperatives, and many other possibilities); and avoid waste.

When I water the plants, I always reuse rainwater, to save money and avoid leaving that water still, reusing it whenever I need to.

At night, if not necessary, I avoid turning on fans or air conditioners, opening only the window to enjoy the fresh air that nature provides us.

Finally, I turn off all the lights so that I can have a good and pleasant night's sleep and with a clear conscience for contributing and playing my role as a citizen to the environment.

The simple fact of separating your garbage will directly affect the collection process of the collectors who live by the search for materials that can be recycled and sold to institutions that work with recycling, in addition to guaranteeing the safety of them and the workers who handle their garbage. somehow, from the janitor of your condominium to the street sweeper who puts the garbage in the truck

MILES ROUTINE

Alice Gomes de Melo

Miles is a little boy who is very aware of how his acts affect the planet. For this reason, he always starts his day helping: As soon as he wakes up, he brushes his teeth and while he is brushing, turns off the tap. The same thing happens when he's going to take a shower. During the shower, while washing, Miles turns off the shower. With this gap in addition to helping the planet by helping the water taxes at the end of the month. Daily, when he goes to school, Miles prefers to take public transport because this way there are fewer polluting gases going into the atmosphere. The time he spends at Miles school also takes the opportunity to raise awareness among his classmates, after all, the more people know the importance of helping the planet, the better. Arriving home he sees his mother reusing the water from the washing machine for other purposes, it is remarkable that Miles has always had an example indoors. In addition to these little things, Miles and his family practice 3Rs:

- Reduce. But reduce what? Reduce the waste we manufacture daily and reduce the emission of polluting gases.

- Reuse. Reusing is giving another utility to something that would be thrown away.

- Recycle. Recycling is giving life to some product that can no longer be used, turning it into something else.

Did you see it? We can't let our home die, it's our duty to fight for it!

ANA'S ROUTINE

Grazielle Ceccatte
Júlia Caroline dos S. Souza

It's Monday, at 5:30, Ana gets up and tidies up her room, after that, she takes a quick shower using water only when necessary, gets dressed, and checks her E-mail, as it was her choice to receiving letters from your grandmother this way and not wasting paper. At 6:30 she takes her reusable bottle and her bike and go to school, getting there, chats with her friends, until the bell rings and then she starts her studies.

At 9:15, recess starts, she eats an apple and drinks a juice with her recyclable straw and then separates the garbage from her food. At the end of recess, she returns to the classroom and struggles to carry out her activities. At the end of the class, she always goes home by bicycle, arriving home, she has lunch and eats until the plate has no traces of food, washes her plate, saving water and soon after, she takes care of her little plant that is located in the window of his room and carefully observes the amount of water, so as not to throw it away.

She spends the afternoon outdoors to save light, reading books on his laptop, and also carries out his weekly activities. That day his mother asked her to walk to some stores to buy sustainable clothes, and off they went, having fun all the way, and on the way back, bringing a good amount of clothes. Coming home it's always time to separate the clothes, donate old clothes, and that she wouldn't use anymore. At dusk Ana and her family stay in front of their house, watching people, and drinking a bottle of water to replenish their energy. At 20:30, she always decides to plan her next day in a spreadsheet on her cell phone. Right after that, she usually takes a shower, lies down on her bed, and sleeps to get ready for the next day.

AMANDA AND CAROLINA' SUSTAINABLE ROUTINE

Carollina Castilhos

Amanda G. Meireles de Souza

Each one usually wakes up at 5:30 in the morning, then we get ready and go to school. We study at Federal Institute of Rondônia, Vilhena Campus. We are the same age which is 16 years old. Now, each one will tell a little about our routine.

Hi my name is Carolina and my routine after school is basically after getting home from school I have lunch. Sometimes, later, I usually throw the trash away, always reusing what I can, and separating each trash in its proper place. I consider it's very important. In the afternoon I usually go for a bike ride. By the way, people should use bikes more, and leave cars aside, this would help a lot to reduce gas emissions that harm the environment and even ourselves.

Hi, I'm Amanda and I'm going to talk a little about my routine, Well, I study in the morning, and when I get home I usually have lunch. I'm trying to reduce meat consumption to have a more sustainable routine. In the afternoon I take care of a baby. Then I usually walk my dogs, when I get home I usually separate the garbage, so that the organics stay in place and the plastics too, everyone should do that, the environment thanks you. I also usually make arts with products that can be reused.

LITTLE LEO'S ROUTINE

Rayssa Monteiro Gomes

Leo is a boy who loves, protects and defends nature, and with the help of his mother he does everything possible in all his actions to do the least harm to nature. so every morning as soon as he wakes up, he brushes his teeth with his ecological brush made of bamboo.

Then, he eats fruit bought at the open market or produced at the family's farm, and Leo helps with all the food. he takes the left-over food and puts it in the compost box, to become fertilizer and they deliver it to Paulo, who he has a small farm and grows fruits and vegetables. And the water that Maria Leticia, Léo's mother uses to wash clothes, he washes the area and the sidewalks of his house with her, Leo waters the plants in the backyard every afternoon, and besides that most of his clothes are bought in Brexós, as it prolongs the life cycle of clothes and, above all, saves money. Actions like this make a total difference to nature and people, if everyone thought the same way as our little big Leo our world would be very different.

MY ROUTINE

Geovanna Rocha

Most people live stressful lives, but with a little effort, you can make your daily routines and have power more effective. So I'm going to talk a little about my daily life.

Hello, my name is Geovanna Rocha, I'm from 2°A high school student, from the Technical course in informatics. I'm going to talk a little about my daily life. I wake up at 6:00, get ready and have breakfast, at 6:35 I leave the house, arrive school at 6:45 or 6:48. My classes start at 7:10 am and 11:30 am to 11:35 am end.

After school I go home and do my homework, and then do my coursework and my assignments. This course is to prepare me for the Enem. After all that, if I have time, I watch a live or watch a video on YouTube, stay until 1:30 am and then go to sleep.

And this happens every day, but there are days when I'm very tired and I go to sleep. On weekends I catch up on chores and play games with a friend of mine, she started

college now and I only study so that's why we play on weekends and every now and then I hang out with some of my friends and hang out with my cat.

YASMIN'S DAILY ROUTINE

Kauã Gabriel Peres Cruz

Yasmin wakes up at 6am, makes the bed, goes to the bathroom and takes a quick shower so as not to waste too much water. Then, he gets dressed in her uniform and shoes, then she goes to breakfast and then goes to brush her teeth, leaves the house and goes to school by bicycle and gets there at 6:50 am. Her class starts at 7:05 am she stays at school all morning until 11:15 am, from where she leaves and gets home. At 11:45 am, she has lunch, washes her dishes always turning off the faucet to don't waste water and rest playing on your cell phone. Arriving at 1:30 pm she does her school homework and at 3 pm she helps her mother to clean the house and throw the garbage in the appropriate recycling bins.

At 4 pm she goes out to play with friends on the street, when it is 5:30 pm she gets home and she makes her dinner always avoiding wasting food and then she goes to draw what is something she loves to do, and she draws until 8:30 pm. Then she takes a shower, brushes her teeth, packs her materials for the next day and watches TV until at 9:30 pm and then she lies in her bed and sleeps for her next day.

LITTLE ACTIONS CAN HELP

Sanderson de Sousa Mantha

Hi, I'm little Sanderson. I'm a big fan of nature. I wake up to water my vegetables like mint etc. Here in my house, we use sunlight to illuminate the area with transparent roof. When my mother washes clothes we use water to wash the opened area of the house.

I do recycling like gathering cans or separating the garbage from recyclable or organic. I also use organic garbage as fertilizer like banana eggshells for my plants. When I need to go out I prefer to go by bike than by motorcycle because in addition to saving gas.

I help the environment, I helped 2 times with food and book deliveries to the rubber tapping community plants called Pau Brasil at the school in the city of Seringueras and helped to plant banana trees and fruit trees and that IFRO school. These are little actions that can make difference to future generations.

SMALL SUSTAINABLE ACTIONS

Iago Davanse de Freitas

My name is Iago and I'm going to talk about small sustainable actions that I do throughout my day.

When I wake up I help my father to water the plants and grass in the garden, whenever possible I separate the garbage.

I always try not to exaggerate the waste of water and I always try to go to places with a bicycle because it does not harm the environment.

In the place where I live there is a little field and always with the help of my father we always plant some trees so that in the future there will be plenty of shade and fruits

AMY'S ROUTINE

Isabelly Silva Aguiar Andrade

Amy wakes up everyday to go to work, when waking up Amy brushes her teeth and takes a shower, she turns off the shower or the sink when not using the water, she avoids using water too long to avoid wasting water.

Amy commutes to work by bicycle, as she lives close to the service and thus helps the environment.

At lunch, Amy always takes a lunch box from her house with the right amount of food to avoid waste, she avoids using plastics, such as straws or packaging for industrialized products.

In the afternoon, when Amy is cleaning her house, she reuses the water from the clothes machine, she collects the garbage and separates it into recyclables or not.

When Amy goes to sleep, she turns off all the lights in the house and unplugs some appliances she won't be using. So Amy ends another day in her life

LITTLE JAMES ROUTINE

Anny Gabriely de Jesus Moreira

Little James wakes up on Monday morning to get ready to go to school. After he makes his bed, he goes to the bathroom, he starts brushing his teeth and while brushing he turns off the faucet when he is not using it.

After he finishes, he enters the shower to take his shower, also turning off the shower while not using it, when he is soaping himself he closes the shower. He gets dressed, putting on his uniform and getting his material. James rides his bike to his school. His family avoids using that car as it causes a lot of pollution.

In the afternoon after he gets home from school, he takes his dog, Bob, to the playground and then come back home before nightfall.

When he arrives he waters the garden and the vegetable, they produce much of their food in their own home.

At home James helps his mother, washing the dishes, while James washes the dishes, he turns off the faucet. When night comes, James and his family leave on only the lights that are necessary, the ones in the rooms that no one stays on. off. James and his family always do their best to help the planet, always saving, we have to take care of our planet.

JULIA'S ROUTINE

Ana Rocha

Julia woke up in the morning to walk with her parents and their little dog named Ben. They always walked in nature paying attention to everything around them. When she got home, she went to shower, and then helped her mother make breakfast.

After having breakfast, she went for a walk around the backyard of her house, and started taking care of her plants, Julia had all kinds of plants and loved taking care of each one of them.

Julia every time she went to the restaurant she took a reusable straw, she ate less meat. Her mother rode her bike to work instead of driving.

On Mondays they didn't eat meat, and to wash floors, countertops, dishes they used vinegar and soap instead of aggressive chemicals.

MY SUSTAINABLE ROUTINE

Patrick Gabriel de Paula

My routine starts from the moment I get up, right after I leave my room, I brush my teeth and start doing my chores, on school days I brush my teeth always remembering to take care of the waste of water, I put food for my animals and start getting ready to go out first thing in the morning, I come to school, study, go back to my house and right after lunch I start my daily chores, cleaning the house, yard, furniture.

Anyway, the basics, I finish everything around six in the afternoon, I take my shower and go play/watch something on my computer, later I have dinner and take care of my animals again, I put water and food and then I go back to my room and stay there until late to sleep.

THE SUSTAINABLE ROUTINE OF LITTLE KAREN

Kauã Fernandes

In the morning, little Karen eats her vegan breakfast with no animal byproducts. He brushes his teeth, after which he tweets making people aware of the vegan cause and sustainable acts. At 9:00 am she goes out for a bike ride and also to make people aware of how cars pollute the environment.

Little Karen always uses paper straws when she goes to lunch, and she always warns people: "turtles don't survive attacks from plastic straws". In the afternoon snack, little Karen eats some fruit before doing her daily exercises.

Every afternoon Karen sorts her trash properly before she takes it out for collection. She always goes to demonstrations and movements supporting the animal cause during the afternoon.

At night, little Karen usually reads and plays her instrument before dinner. At dinner she eats with her family, brushes her teeth and goes to sleep.

CREATING A BETTER WORLD

Alêssa Silva Peçanha

Geovanna Duarte Lima

Aurora and Cecília are friends and they study together. Whenever it gets dark they have coffee and talk about human actions and how bad they are for the world, then Aurora and Cecília realize that they also need to change their attitudes.

Cecília is fifteen, she cares about nature and the planet´s situation.

Every day, she wakes up and thinks about how to act to have a more sustainable life. She is a very conscious girl. On Monday morning, Cecília organizes her closet and chooses some unused clothes to donate to those people in need or to a charity shop. At school, Cecília usually avoids wasting paper and she advises your friends to do the same. She also prints off documents only when it is extremely necessary. On weekends she keeps recycling and reusing what is being sent to the trash, of course some things can not be recycled but there are a lot that can be. With creativity, she changes her whole space.

Aurora is a sixteen-year-old girl and has always wasted food, but this changed because now she makes composts. Sometimes she leaves the lights on, at her house, but it is not common anymore. In the morning, she doesn't turn on any lights, and just when it's necessary she turns it on. She also saves water when she brushes her teeth and moreover, doesn't take too long in the shower. She often uses soap and vinegar instead of harsh chemical products to clean floors and countertops at home, and she doesn't throw a lot of water away. She rarely uses cars for transportation, she always rides her bike to school. Aurora eats more vegetables and greens than meat. When she goes to the market, she always reuses her bags. She always buys some plants for her house.

Aurora and Cecília share the same ideal of the world, that is: a clean, nice and livable environment as it was before the polluting actions of men. Together they plant a tree every week and they are planning speeches about environmental education for the school's external society.

URBAN UTOPIA

Luiz Fernando

Vinícius Almeida Moraes

Hi everyone, my name is **Frank** and I live in the city of **Chicago**. Today I'm going to tell the story of how I decided to revolutionize my way of living.

I used to wake up every day with my head spinning, unwilling to build new ideas and new thoughts. I felt like a puppet of the environment in which I lived. Always waking up with that city penetrated by darkness, noise and visual pollution, smoke and a hot wind.

It was then that I decided not to put up with it any longer and start a revolution of my own. I made a radical change, threw away everything that was bad for me, drinks, treats and all that stuff. I started to fall in love with the contact with more alive, natural things and without contact with industrialized ones. I moved to a part of the city closer to the countryside, where I adopted several animals that gave me the joy that was not provided in the noisy solitude of a place full of people.

It seems comical to say that I felt alone even though I was surrounded by many people and things that gave me instant pleasure, but it was something that was simply not created for me. I disconnected from the urban environment and surrounded myself with what our planet has to offer, such as harvesting from my own backyard what I planted, swimming in a river full of fish, running on distant green lawns and many other wonderful things that life provides. In the near future, who knows, I intend to take my message to all those who suffer from the same problem I suffered in the past. It's not difficult to maintain, but starting is like going into a dark hole, not knowing what you'll find on the other side, but the light at the end of the tunnel is most of the time right.

JOSEPH'S NOT VERY STRANGE ROUTINE

EduardoTartas

Yuri Ribeiro Zetoles

Sustainable routine what is it?

We will demonstrate using a routine of a fictional character, he is called Joseph. Joseph always starts his morning by making his bed right after waking up, then he goes to the bathroom to take a quick shower and brush his teeth, always aiming to save water, quickly he changes, then he goes to have his breakfast which is completely vegan because Joseph supports animal causes.

Joseph goes to school first thing in the morning by walking or cycling, because he doesn't use cars that emit carbon into the atmosphere, when he gets to school he doesn't use disposable cups because he has a reusable and recycled bottle because he avoids plastic disposal as much as possible in the environment, at lunchtime he invites his schoolmates to try a vegan lunch, to show them that it is not necessary to eat meat in meals to be nutritious.

In the afternoon Joseph comes. home from school and starts getting ready to go volunteer at an NGO, he gets ready and takes his bike and goes to the NGO where he helps homeless animals, giving them food, water and finding a home for them., it also helps in the reforestation of deforested areas and makes people aware of the risks of deforestation and global warming.

At night Joseph always does his chores with recycled paper so he doesn't contribute to deforestation, he also helps his parents with housework and goes to bed early to have a healthy routine and also not use too much energy.

"If you face the abyss they call the "future", and manage not to fall into it, you will never have "downs" in your life again."

LASTING LIFE

João Vitor Alves de Melo

João Vitor practices a sustainable routine. First he knows that daily actions help life on earth to have more support, not only in the environment but in health and the economy.

Immediately, João buys reusable packaging, stops using paper towels and replaces them with a dish towels, buys durable bags to avoid using disposables, uses bar soap instead of liquid, reuses tea bags to neutralize odor, and uses batteries rechargeable.

João doesn't buy plastic straws, has glass jars instead of plastic, does not brush with a plastic brush, makes coffee with a cloth filter, washes clothes with cold water, sometimes doesn't drink water from disposable battles, buys reusable cups, no eats a lot of processed foods and takes a daily shower In a water-saving shower.

These are the sustainability habits that João practices every day, if possible.

ROMEO'S SIMPLE SUSTAINABLE ROUTINE

Maxsuel Fernandes Bispo

Well, let me introduce you Romeo. He is self-employed, lives in the city of Vilhena and he daily practices a sustainable routine, as he believes that this can positively affect the environment around him.

He always wakes up at 06:00 in the morning, this helps him to have more energy during the day. After getting up, he goes to the bathroom to brush his teeth. Thinking of a sustainable action, while Romeo brushes his teeth he turns off the faucet, remaining open only when rinsing his mouth to avoid wasting water.

During breakfast Romeo prepares a simple sandwich with whole meal bread and salad. He always takes advantage of every part of the vegetables, without waste because he believes in the benefit of vitamins in food.

Daily he cleans the house. It does not make use of electrical equipment, as using such equipment causes an increase in electrical energy consumption. Thus, opting for other means of cleaning, such as, the use of the good old broom, squeegee, bucket and a floor cloth. At the end of the cleaning, he always separates the garbage by disposing of it in the proper place, organic materials such as food scraps are used in the composting process, dry materials such as plastics are discarded in the selective collection, which takes place once a week in the neighborhood where he lives.

Sometimes, at the end of the day, Romeo goes to the local open market with his bicycle, in this exchange of the car for the bicycle he practices a physical activity, saves fuel and doesn't waste time with traffic jams, because the sustainable economy is also about health.

Do you remember the sandwich he made for breakfast? Well, it's at Dona Tereza's tent that he finds fresh vegetables, without any kind of pesticide use. With this action, he promotes the local economy, giving preference to local production, where the final price is more inviting, as the expenses with transport and storage decrease.

During his purchases he is careful not to buy anything on impulse, buying only what is necessary. This practice helps Romeo save more money.

Right after his shopping, he makes the trip to his house again. When organizing his purchases, he reuses packaging, such as plastic bottles, for example, he turns them into potted plants for his new mini cactus seedlings, thus reducing the waste produced at Romeo's house.

Finally, he takes a quick hot shower and goes to bed, because tomorrow his simple sustainable routine will repeat itself, this time with new challenges.

DAILY LIFE OF A CONSCIOUS CITIZEN: PRESERVING TODAY

Beatriz Archanjo Moura

We will talk about the day of Ana who is a student, she starts the day waking up every morning at 5:00 to be able to get ready to go to school, soon after she goes to shower and brushes her teeth always trying to consume as little water as possible. At 06:39 she has breakfast with healthy food and eat the ideal amount so there is no food waste. Right after breakfast she goes to school by bike.

In her school that calls itself, Ana along with her classmates will do a different task which is to plant trees and flowers in parks and public places that need green areas, and also do the cleaning of a park that was dirty, taking bags, bottles, glasses, soda cans, papers and other types of garbage in its proper place. In her school it is common to have these types of activity, because it helps to show students that everyone needs to take care of the planet in which we live even in the smallest things we do in our daily life.

After school Ana usually goes to the market to buy vegetables and other things to prepare lunch. She doesn't like to use plastic bags because they are made of oil and they last hundreds of years, and they spend all this time causing damage to natural habitats and killing animals because they eat the bag thinking it is food, then she always carry with her Eco bag. After arriving home and having lunch Ana goes to the gym at 14:15 on foot and back at 15:00, then shl takes a bath always saving water, studies for a few hours, she always has a habit of taking the trash out of the house, breaking the trash for selective collection. Finally, she ends the day with a dinner without the consumption of beef and will rest for the next day.

DAILY ROUTINE

Pedro Henrique Ferreira de Oliveira

Damian starts his day at 6:00 AM. Then he has his breakfast and brush his teeth. Damian doesn't like to waste water, so while he brush his teeth, he leaves the faucet off, he does it all the time. At 8:00 AM, he enters the job. Damian is a "sustainable engineer", he does sustainable projects for his city, like container houses, places to put recyclable waste and more. At 12:00 he has his lunch. When he's finished eating, he always puts the garbage in recycling bins.

At 18:00 he backs home. When he's in home he take a shower and has dinner. Damian likes very much watch documentaries and series about the nature, before sleep he read books and listen musics. At 10:00 pm he goes to bed.

DAY BY DAY

Arthur Henrike

Guilherme lives in Vilhena and studies at IFRO in high school, he wakes up at 4 am does physical exercise, and eats breakfast with various fruits and oats.

After breakfast, he brushes his teeth, always using as little water as possible, turning off the tap while brushing his teeth.

During his morning he studies he is very interested in electronics and computers, and he is always looking for something new to learn, that day he wants to learn to control the Arduino after 2 hours of studying he gets ready to go to school. His means of transport is a bicycle and when it is rainy days he goes to school by bus, he arrives at school, today he has Portuguese and mathematics classes. Usually, he has a lot of difficulty in mathematics that he ends up using almost all his effort to learn math by writing down on his computer what he understood about the subject.

At the end of his day he gets home at 5:50 pm, today is garbage day where he takes all the garbage in his house and divides it into categories: glass, paper, metal and plastic, after that he separates his breakfast the next day and leave it in the fridge, tomorrow will be a special day for Guilherme because this is the only time a week he sees his mother, especially this week he visited his mother to pick up his old toys and donate to an orphanage near his house after that Guilherme goes about his normal routine, this is Guilherme's routine

PEH-YAN AND HIS SUSTAINABLE DAILY ROUTINE

Luiz Fernando Gomes

Peh-Yan wakes up at 4:00 am to go for a morning walk with his friend Pah-chin, they meet at Pah-chin's house, then they go to a square that is 1 km from Pah-Chin's house, then they come home again at 5:00, so Peh-Yan makes a nice jelly salad sandwich, and has a glass of fresh juice.

At 5:30 he takes a shower saving as much water as possible. He finishes his shower at 5:45.

At 6:00h he goes to work by bicycle, where he arrives in 15 minutes.

At 7:30 it's time to rest from his work, he takes advantage of this time to clean his desk and put the papers he's already used and he doesn't go use more for recycling, he goes back to work at 8:00.

He gets home at 12:45 pm and goes to clean the house with the water he used to wash clothes, he finishes at 1:00 pm, then he goes to prepare his lunch, free of red meats and very fatty foods, he has lunch at 14:00h.

At 2:30 pm he will wash the dirty dishes, saving as much water as possible.

At 3:00 pm he goes to read books, at 4:00 pm he sweeps the yard and sees the materials that can be recycled and, if he can, he recycles them.

At 5:00 pm he goes to the market to shop using his ecological bag, which is more expensive than plastic bags, but can be used many times and does not harm the environment.

At 18:00 he gets home and goes to pack the groceries and then he goes to meet his friend Pah-Chin so they can play basketball, which is their favorite sport, at 19:30h he comes home and drinks bath, saving water.

At 20:00 he prepares his dinner, at dinner he likes to eat only healthy foods, after eating he washes the dishes, and at 21:00 he goes to sleep to wake up early the next day. And this is Peh-Yan's routine, what do you think?, are we going to do like him and try to have a life as sustainable as possible?

ELLEN'S SOLUTIONS

Emily Santana

Ellen's dream is to help stop the pollution of the world, she is only 10 years old and as soon as she wakes up she observes everything during her day and chooses something that pollutes the environment for the night she does her best to prevent such a thing from continuing. advancing. Today Ellen woke up at 6:30 as usual, had her breakfast and got ready to go to school, Ellen rides her bike to school alone, since she lives just 4 blocks before school, when she arrives at school Ellen comes across a lot of different people, they are environmentally conscious and had gone to school to make students aware of how harmful to the environment improper waste disposal is.

Upon learning about this, Ellen decides that this will be the problem she will solve today. After school, Ellen goes back to her house for lunch and in the afternoon she has swimming class. When she finishes these tasks and goes back to her house Ellen thinks about how she can solve this big problem, researching on youtube about.

So, she decides to send awareness messages and send it over the internet to everyone she has told, reminding them that there are selective collection wastes and that no one will be harmed more than the their own generation.

PAULO'S SUSTAINABLE ACTION

Amanda Rafaela Freitas de Mathias

Taline Rodrigues França

Paulo is a 30-year-old man who teaches mathematics at a university, loves sustainability and always lives it in his daily life. When he wakes up every day, Paulo avoids wasting water, it doesn't take long to take a shower, in addition, he washes the car with a bucket of water.

On Tuesdays and Thursdays he rides his bike to work. As this process is constantly changing, he avoids daily driving. On certain weekends he hangs out with his co-workers, where he uses his reusable straw.

Paulo every year plants trees in his city's parks, participates in projects, saves energy and water to ensure good sustainability.

At 23 years old, Paulo still doesn't understand all types of sustainability very well, but every day he studies and improves his lifestyle to generate a better future for him and for helping people.

CAIO THE DOCTOR

Kauany Godinho

Rittielly Mangarotti

Caio is a 27 years old doctor and lives alone. He waks up at 5am when he's not duty. Caio takes everyone's health and well-being very seriously, and tries to help in any way possible by doing volunter work for those who can't afford a medical appointment and is very strict.

When it comes to the 17 sustainable goals and your goal will help the most using these ways. His way of going to work is quite unusual for some workers, as he always rides a bike regardless of the place and says that it will help him not to pollute the air so much being his hobby.

Even though he thinks that he does little, he always gives his best because he cares about everything that involves the planet and the next.

He has always thinks he can do more than that. This is Caio's routine being sustainable!

RECICLE IS NECESSARY

Thais Cristina Noemerg Crispim

Julia Dourado Silva

Rural producer Elaine, has always been concerned about the negative impacts caused by waste discarded without any treatment on the environment. She believes that it is the duty of all citizens to help take care of the planet and not just the public power.To try to soften a little,she decided to do her part and started to recycle the waste of her property, separating organic waste from solid waste.

When he started to make these separations of the tailings, he realized that there was no waste, but products that cold be reused in several ways.

The organic waste went to a compost bin and turned into very rich fertilizers that are very good for the plants, and the recyclables cooperative to become new products.

Like Elaine, all citizens have a duty to take care of the planet, avoiding excessive consumption, and recycle all the waste they produce, because there is no garbage, everything can be transformed. If everyone does their part, the planet will be better

LITTLE CHANGES FOR A BETTER FUTURE

Ester Jenifer Cândia Boa Ventura

I started a sustainable routine not long ago.

There are times that in the rush of the day, I forget to take my reusable cutlery and glass. Since I learned about how to reduce the production of garbage in my house. It turned out that I managed to save a lot of money, apparently, changing the dishwashing sponge for vegetable loofah doesn't only help the environment. Garbage separation also works perfectly and both are happy: Not mixing plastic with organic and washing the packaging and bottles makes me feel less disgusted about touching the garbage. Using glass containers instead of plastic, and having some plants is much more aesthetic for my home.

My wardrobe organization has reached another level; I donated clothes that no longer fit, in addition to helping others, I feel much happier to see everything tidy.

Anyway, I still have a lot of things to add to this new lifestyle, but I think that no longer not only the planet will get a little better, but also will I.

NEW AGES, NEW CHALLENGES

Fernanda Clemente Del Barco

One subject that is being talked about a lot these days is sustainability. Nowadays, the world is looking for more sustainable alternatives so that the planet can be more preserved.

Sometimes, a small action, such as turning off the tap while brushing your teeth, can make a huge difference on the planet. Sustainable development aims to preserve the planet and meet human needs. A natural resource if exploited in a sustainable way will last forever and our future generations will also be able to enjoy it.

However, we must all do our part to have a more preserved planet. Sustainability is divided into three principles: the social, which encompasses people and their living conditions; the environmental, which refers to natural resources and how they are being used by society; and the economic, which relates to the production, distribution, and consumption of things.

Something I don't understand is how society can close its eyes to something that is so important to be discussed and worked on nowadays. If people would simply reduce the amount of water, recycle, among other examples of sustainable actions, the planet would become such a better place to live.

WE NEED DO SOMETHING

Grazziely Fernanda Boing Santos

To have a healthy and happy life, we need to take care of our habits. Save energy, save water, don't waste food, reuse and recycle. We can place plants in rooms in the house to improve the quality of the air we breathe, walk, or ride a bike to school or to work.

We can start with something simple, for example, mothers, they always love flowers. They don't forget to water their plants. We need to change our mindset, start thinking more about the world we live in. The earth is screaming every day. It doesn't have anyone to help to take care of the planet.

We need do something to change this, why we don't save more energy, or eat less meat? We can plant trees in our backyard, or reuse the water we use. Like the water we wash clothes, we can use it to clean the house, we all know that the smell of fabric softener is wonderful. I know this might sound a little weird, but why can't we try? I want to do something to help, I don't like to think that I'm not doing anything for the planet, how about you?

There are some healthy habits we can adopt. Save water when washing your face, brushing your teeth and taking a shower. When we serve ourselves during meals, avoid unnecessary waste. Change the means of transportation for a walk or go by bicycle. Use cloth bags instead of plastic ones, avoid the use of disposable utensils, avoid the waste of paper.

Do you want to try? We may get it together!

A SUSTAINABLE ROUTINE

Hynaê Messias Pires

On school days, I usually wake up at 5:30 am to shower, get dressed, and water my house plants. I walk to the bus stop with my friend, and we take the bus to school together.

When I get to school, I usually fill my reusable bottle at the drinking fountain.

There are some days that I have to have lunch at school because it's a full time study. As I'm a vegetarian, I prefer to take food from home to school. But there are some days when I have to eat at the school restaurant, on those occasions I choose to eat healthier. But I have few options.

On the days when I don't have afternoon classes, I like to go to the street fair. There, I buy some vegetables and fruits and put them all in my reusable bag. I'm responsible for cleaning the house for a few days in the week. I typically use a mixture of vinegar and baking soda to clean some things. It is eco-friendly and healthy because chemicals have a strong smell and are bad for health.

At the end of the week, I do my homework and separate all the papers that can be recycled, because every two months I make recycled paper, which I use to draw, make notebooks and print.

I don't keep a fixed routine, because I know that being restricted is ungood for me. So I try to do things in my week that are good for the environment and me, because I live in it and I want to keep it preserved for future generations.

TIPS FOR IMPROVING QUALITY OF LIFE

Maria Destri

The moment we faced about social isolation make some people rethought about different attitudes. For example, on how people can have a more sustainable routine and life.

When we were at home all day and couldn't often go to the grocery store, the challenge was to reduce waste and still keep the food fresh.

In addition, to the kitchen, conscious consumption also approaches toys that are made of plastic. Avoid wasting water and food. Cooking only what is necessary collaborates with the Environment, because there will be no need to throw away the "leftovers" of food. Avoid buying bottled water if you have sewage treatment at home. Thus, you save money and also mineral resources.

Give preference to products that use batteries or rechargeable batteries. The exaggerated consumption of non-reusable batteries causes a great environmental impact, as they contain chemical elements.

Enjoy the sunlight. Depending on the layout of your home, it is possible to use sunlight until the beginning of the night. In addition, replacing normal light bulbs with energy-saving light bulbs will spend less energy and will last longer.

Use your feet (or bikes) more. Use more public transport or bicycle. walk more! If you walk more, in addition to helping the planet, it will also be good for your health.

Don't waste paper. Also take the opportunity to save on prints.

Separate the trash. Maybe this is one of the simplest tips, but not everyone does it. Cultivate green areas, the greener the better. These are simple attitudes, but they make all the difference to our Planet.

SMALL STEPS FOR CHANGES

Sarah Kimberly D. C. Costa

My routine starts at 5:20 in the morning.

One of the first things I do is my hygiene, starting with my teeth and ending with the shower. But when it comes to sustainability, I always try to save water as best I can, not overusing it. Then, I get dressed and I rarely drink coffee.

My routine is always the same, especially when I have extra shift sat school. Another fact that can be considered sustainable is to always properly separate waste and avoid those that cause extreme damage to nature, such as always avoiding the use of straws.

I don't know Much about sustainability, I should learn more, but maybe the fact of using vegetables and fruit peels in gardens instead of throwing them in the trash are something considerable, and it's something I do in my daily life.

GIRL MEETS THE WORLD

Emilly Windila

With the sound of the alarm clock, Ashley wakes up in an autumn morning. She gets up from bed, takes a bath, and brushes her teeth. It's nice Wednesday of weather, Ashley likes Wednesdays a lot because her restaurant is shorter. She leaves home 30 minutes before her shift starts. Ashley is a restaurant manager. She has been working in this field for four years since she left university.

Ashley has the habit of always passing by the caffeteria before work to buy a coffee, near the restaurant there is always a gentleman asking for Money. She always tries to help him. But that day she noticed that this gentleman was not there. She freaked out, because, he always sits in front of the restaurant.

She went to work and stayed there until her shift time was over. On her way back, she passes by the restaurant, but the gentleman she used to help was not there.

The next day, with the sound of the alarm clock, Ashley wakes up, she gets up from bed, takes a showers, brushes her teeth and goes to work. She was hoping to find the man, she always helps, but that morning there was a woman sitting in that gentleman's seat.

She became surprised. The woman was the lord's wife who was sat there. Ashley addressed to the woman and asked what had happened to the gentleman that she always helped. The woman replies to her cold and empty, "he died, for him always sat there, regardless of sun or rain and ended up that he got sick".

Ashley shocked by the news, comes home. She grieves, but she doesn't let her know that gentleman had a family she decides to do her best to help. She as a manager, had enough money to help that Family. So she decides to rent a house to them and gives a job in the correctional to the woman who had just lost her husband.

In the long days, after these events, she realizes that her family begins to stabilize financially and that she no longer needs to ask for money. After Ashley's deeds, she decides to go deeper and discovers that much of the world's population lives in the same condition that that family was in. So with the help of some friends of the manager, she decides to open an NGO where it aims to help homeless families who need help financially.

Always at the time going to work friends and other people who wish to help leads food help with donations, with these helps she began to devote only to NGO. The truth that people only care about certain events run away from their routine, we can all help without getting out of their routine, on the contrary, make it your routine.

GOOD PRACTICES FOR THE FUTURE

Kerlon Alves Moreira Borrher

Maria wakes up and does her needs. One of them is having breakfast, as she doesn't like coffee she drinks a natural orange juice.

Soon after, she starts doing her walk whose walk she does twice a day for not having time on mondays and thursdays, she only goes for a walk in the morning. Every now and then, in the middle of her walk, she stops by a library that is about two blocks from her house, to guarantee her monthly studies.

On her way to work, she stops by a flower shop and picks up some seeds to plant in her college garden where she studies at night. Once a year, she participates in a science fair where she presents sustainability projects to her colleagues and to her teachers.

One thing she practices every weekend is her cooking as she doesn't like to order food. Another thing she does on weekends is design projects to reuse water to improve the environment. To save energy Maria tries to keep most appliances that are not being used at the moment turned off.

She also doesn't use plastic to preserve the environment that in her view is already well polluted. Her main means of transportation is her bicycle which she uses to go to places far from her home.

Mindset on sustainability

GUSTAVO ANDRÉ DIAS DOS SANTOS

Gabrielly Menezes

As we can see below, Jhon is a person with a sustainable mindset and lifestyle, it wasn't always like that, but he soon decided to change his mindset and became the person he is today.

Jhon doesn't like to waste food, everything that he doesn't use in his food, he turns into fertilizer for his vegetable garden, which is behind his house, there he plants and grows fruits, vegetables and legumes, because according to him, these such foods sold in supermarkets are not fresh.

Everywhere he has to go out, Jhon goes by bicycle and takes his little bottle of water. So, he doesn't use plastic glasses. Cycling is his only and preferred means of transportation, what good would it be to practice all these customs for a better world and ride a bike right?!

As Jhon aims to reduce the use of paper in his life, he prefers to receive receipts from stores, letters, documents and other things like that, by email, he only prints something if he really needs it. Jhon is a 30-year-old American boy who chose to have a healthier life, thus following dozens of rules to achieve his goal.

He is a healthy person, loves to exercise and eats well. During the week, he wakes up at 6 am, meditates and then has his breakfast. After all is done, he likes to go for a walk and when he comes back, he likes to swim in his pool.

To maintain a healthy and balanced diet, Jhon chose to eliminate red meat from his life.

In order to help eradicate hunger in the world, Jhon prefers to shop in responsible companies, which help charities that welcome people in vulnerable situations, and in places where trade is fair and reliable. When shopping, Jhon always takes his sustainable bag, so he doesn't accept plastic or paper bags provided by the company, as he doesn't like and doesn't agree with their use.

Every month, Jhon helps a different charity and donates items he no longer uses.